# WALK ON WATER

PAUL BREER Ph.D.

**Quantum Discovery**
A LITERARY AGENCY

Frontcover painting (Metamorphosis) is by Varouj Hairabedian

Library of Congress Control Number: 2025918269

ISBN
979-8-89641-101-7 (Paperback)
979-8-89641-102-4 (eBook)

*Despite its small size, the human brain is one of the most powerful devices on earth. Unfortunately, we have wasted its power in a futile attempt to promote and defend an entity that exists only in our imagination. That entity is the self.*

*Dispelling the self-illusion will change the way we identify who and what we are. No longer will we see ourselves as "masters of our fate, captains of our soul" but as powerful and elegant organisms guided by genes and upbringing, less self-absorbed, more caring and humble, yet capable of feats traditionally reserved for the gods.*

# Table of Contents

# *Introduction*

Traditionally miracles have been defined in two different ways, and often as both. On the one hand they are thought to be the work of a supernatural entity, typically a god or goddess who intervenes on behalf of humans here on earth. On the other, they are considered miraculous because they violate one of the physical laws that govern the universe. But both of these definitions are under attack, whether from philosophy or science. As the world moves towards a more secular view of reality, the existence of divine beings in heaven who every now and then take pity on humans and perform a miracle for us is coming into question. At the same time, our understanding of the physical universe has broadened to the point where the traditional view of human existence as a collection of separate selves who are basically unconnected to each other no longer fits the results from scientific research. Taken together these changes in our thinking about miracles are shaking our faith in what was once considered inviolable truth.

Writers who share that position can differ in the nature of their doubts. Some believe that the events purported to be miraculous are nothing more than the product of wishful thinking. In other words, they never happened. Others believe that at least some of them may have happened but dispute the view that they were caused by gods and goddesses or represent a violation of basic physical principles. Recent research in the fields of quantum physics and parapsychology suggests that the latter view is more likely correct. If true, it leaves open the possibility that it is not the gods but we human beings who possess the capacity to make miracles. In The Conscious Universe Dean Radin makes a case for this view.

> At a minimum, genuine psi [psychic phenomena] suggests that what science presently knows about the nature of the universe is seriously incomplete, that the capabilities and limitations of human potential have been vastly underestimated, that beliefs about the strict separation of objective and subjective are almost certainly incorrect, and that *some* "miracles" previously attributed to religious or supernatural sources may instead be caused by extraordinary capabilities of human consciousness.[i]

While still contested by many skeptics, research in psi is creating possibilities once considered outside the limits of science-based thought. As that body of research grows, more of us are willing to entertain the argument that so-called miracles may not be miraculous after all, i.e., they neither originate in heaven nor do they represent violations of basic physical laws, which suggests that these rare and wondrous events may be performed not by unseen entities in the heavens but by humans operating within what we now know to be an "entangled" universe.

Although we may wish to deny it, the model of the universe constructed by scientists of the 17th and 18th centuries (known as "classical physics") is the same model of the world that most of us cling to today. Since there was no place for miracles in that mechanistic model, the only alternative for our predecessors was to attribute miracles to the gods who are unencumbered by the constraints of cause and effect. As our scientific and philosophical thinking has changed radically since 1700, the way we interpret miracles has undergone its own metamorphosis. While at least some of the events remain believable, we are more inclined today to view them through the lens of 21$^{st}$ century physics where they can be treated as phenomena rather than gifts of the gods—still remarkable but no longer originating in a world apart from our own.

# 1

## *Validity*

I think we can safely assume that very few of the miracles we hear about are true. Most are either fantasies which when told and retold across generations, morph into real events; others, while containing a grain of truth, are exaggerations of what actually happened. While it is important to keep such caveats in mind, it would be a mistake to discount miracles altogether. In one form or another they appear in just about every culture we know about—and for good reason. By stretching the boundaries of reality, they give hope to those who believe that unseen forces stand ready to provide help when needed, especially when the demands of everyday life seem overwhelming. The continued popularity of miracle stories, even in an age when scientific thinking is widely embraced, is testimony to the urgency of that need.

But the question remains—are any of those stories true? Keep in mind that it takes just one proven miracle to demonstrate that there is more to reality than meets the eye. But even if we can document such a miracle, the question of interpretation remains. The 11th edition of the Meriam-Webster dictionary defines a miracle as "an extraordinary event manifesting divine intervention in human affairs." Does that mean (as all religions insist) that miracles originate in a transcendent realm beyond the reach of human reason? Historically it has meant that exactly, but advances in physics have opened up a new possibility, namely that those same events can be explained in terms of laws (patterns) that operate in the subatomic world. If that's true, it may be more accurate to see such events as natural phenomena that obey modern physical laws even as they violate traditional (17th century) ones. In that case, there would be nothing unnatural about those events, even the most extraordinary ones—so we should stop referring to them as "miraculous" or use a different word. An alternative solution (one that is adopted here) is to retain the word miracle but confine it to "extraordinary events that violate the laws of *classical physics.*"

There exists a wide range of phenomena that appear to violate the laws of classical physics. Events along that continuum differ in several respects: their rarity, the extent to which physical laws are violated, and the consequences of the event (the extent to which a person's life is changed either consciously or materially as a result). At one end of the continuum there is the common experience of thinking about a specific person just before you receive a call from that person, implying the possibility of telepathic communication between the two parties involved. Further along the continuum is dreaming about a lost object and waking to find it in the exact location displayed in the dream—a case of clairvoyance or what is sometimes called remote viewing. More extraordinary is the act of moving a physical object with your mind or curing someone's cancer through mental effort alone. Then there are those events like levitation or walking on water that constitute a total abrogation of traditional physical laws.

While we in the West have long considered miracles to be the exclusive prerogative of the gods, this is not true of other cultures, Tibet being the primary example. Consider the following statement about miracles from Alexandra David-Neel who spent thirteen years exploring Tibetan culture before publishing her findings in the 1920's.

None in Tibet deny that such events may take place, but no one regards them as miracles, according to the meaning of that term in the West, that is to say as *supernatural* events Indeed, Tibetans do not recognize any supernatural agent. The so-called wonders, they think, are as natural as common daily events and depend on the clever handling of little-known laws and forces...All facts which, in other countries, are considered miraculous or, in any other way, ascribed to the arbitrary interference of beings belonging to other worlds, are considered by Tibetan adepts of the secret lore as psychic phenomena.[ii]

If miracles are rare but perfectly natural phenomena as the Tibetans believe them to be, does it follow that all human beings are potentially capable of performing them? My guess is that they are. If so, why has it not happened? I think the answer is obvious: it requires the ability to sustain a highly concentrated form of consciousness which very few humans have developed. But that, in turn, raises another question: why have so few of the advanced meditators used their gifts to perform miracles? That too is clear, this time from the roshis and swamis themselves: they don't want their adepts to pursue miracles for fear that it will distract them from the more important task of becoming enlightened. And so, it remains true (particularly in the West) that we rarely hear about miracles, especially the authenticated ones. But this may be about to change.

Different types of miracles (all involving use of mind alone):

<u>Telepathy</u>

Reading someone's mind or imparting content to that mind.

<u>Clairvoyance</u>

Locating objects without the use of one's biological eyes.

<u>Remote Healing</u>

Curing someone's disease without use of surgery, medicine or pharmaceuticals.

<u>Psychokinesis</u>

Moving physical objects using nothing but your mind.

<u>Alchemy</u>

Transforming objects into something different.

<u>Defying Gravity</u>

Levitation or walking on water

<u>Changing the Weather</u>

Altering temperature, moisture or gas content of the atmosphere

# 2

## *Miracles and Physics*

Whatever words we use, most people think of miracles as rare events which involve a contradiction of the physical laws governing everyday life. But there is a nagging question hidden in that way of thinking. What, for example, if the so-called "known" laws of physics no longer hold true; what if they are found to be only partially true or even contradicted by more recent findings? Keep in mind that the basic physical laws we live by today are pretty much the same ones that Newton, Galileo, and Kepler discovered back in the 17th and 18th centuries. In the intervening time, however, science has moved on, making some of those old principles untenable.

In the early 20th century Albert Einstein turned the old physics on its head with his writings on space, time, and relativity. As dramatic as his writings were, however, they did little to change our view of miracles as contradictions of physical law. That took a few more years. It was when Max Planck solved a problem vexing most physicists at the time (he argued that light took the form of discrete energy packets or *quanta*), that the era of quantum physics was born. And now, more than a century later, that discovery is changing the way we think about lots of things, including miracles. It is forcing us to consider the possibility that extraordinary events like telepathy, clairvoyance, levitation and seeing into the future may be fulfillments rather than violations of physical law. And this shift has far-reaching implications not only for science but for philosophy and religion.

For example, as the new way of thinking sinks in, we may find that it makes more sense to view those miraculous events not as the work of divine agents, but rather as natural expressions of quantum principles. That is dramatic, but what is likely to follow promises to be even more amazing. It suggests that by learning how to work within the principles of the new physics, we humans may be able to perform miracles that were once reserved for the gods.

While quantum assumptions about reality have been accepted by most physicists, there remains a stubborn minority among both scientists and philosophers who cling to the more familiar beliefs of traditional physics. This is understandable given the strangeness of the quantum world in which sub-atomic particles are essentially unpredictable. For example, they move when you try to measure them—with the frustrating result that you can never know for sure where a given particle is or how it reacts to a certain kind of influence. Given the strangeness of this quantum world, if you want to explain how miracles work, you need to re-conceptualize curing a friend's cancer with the mind alone as something more secular, e.g., like "mental influence at a distance."

With that in mind, you might want to start with something familiar and measurable like a brainwave. Of course, the question will immediately arise as to how that brainwave (or more likely a configuration of brainwaves) can not only reach a target person miles away but cure that person's cancer or read her mind without any overt physical intervention. How can a set of brainwaves in person A's mind reach person B who is not physically present—and cure her cancer or read her mind—all without employing tools of any kind or even speaking to her? In short, what role, if any, do brainwaves play in the performance of such extraordinary events?

As some scientists have suggested, brainwaves have the capacity to generate electro-magnetic fields and direct them to alter and re-organize the form of material objects outside the brain. That doesn't mean that they create material objects from scratch; they simply alter what is already there. Stated another way, energy fields form the templates around which matter condenses—change the field and you change matter. And that holds true whether you're trying to change people, zebras, trees, rocks, or even individual cells inside your own body.

This theory assumes, wrongly I think, that the process originates in consciousness where a thought arises, causing neurons in the brain to fire, thereby creating an electro-magnetic field that radiates out through the skull. Based on recent research in the neurosciences, I think it more likely that the process works the other way around—i.e., it originates in the brain—with consciousness serving as what philosopher Thomas Clark calls a "representation" of select aspects of the activity taking place in the brain.[iii]

Assuming that the process works this way, we are still left with a major question, namely how does a configuration of brainwaves (and the energy field it produces) get addressed to a specific target, e.g., a distant friend who is suffering from cancer? As powerful as they might be, brainwaves are no more than oscillating cycles of electricity. While they differ among themselves with respect to frequency (as measured in Hertz) and amplitude (the difference between peaks and valleys), they lack the capacity to aim themselves at a chosen person or a particular problem. But there may be exceptions. There is some evidence that certain brainwaves are typically involved in specific paranormal events; theta, for example, is almost always present in the act of remote healing, first in the healer then in the person being healed. Even there, however, it is hard to believe that an electro-magnetic energy field produced by brainwaves is capable of finding its way on its own to a specific person suffering from a specific problem.

There's another problem here, a serious one. For waves in the brain to form an electric-magnetic energy field outside the brain, they have to first pass through a tough barrier, the human cranium. Most of them don't make it, leaving those that do get through without the concentrated power needed to reach a remote target and heal a serious illness.

The point to all this may become clearer if we switch our attention to a different kind of paranormal event, e.g., clairvoyance. Imagine that your car has been stolen and the police have been unable to locate it. You sit down to meditate, focusing all your mental energy on the missing car and it works. The car is found precisely where it appeared to you in meditation. But how did that happen? Yes, as the owner you knew the make and model of the car as well as its year and color. But do the brainwaves in your head share any of those details? Unlikely. Even if they did, they must be sent to the target, often over great distances. But if that's the case, one of the basic principles in physics comes into play—the law of Diminishing Returns which says that the greater the distance between sender and receiver, the longer the influence will take and the weaker will be the communication, all of which is contradicted by the evidence we have from studies conducted as far back as the 1920's. In those studies (most of them Russian), it made no difference whether the sender (in this case a psychiatrist) and his target (a patient) were in the next room or 1,000 miles apart. The ability of the therapist to awaken his patient from sleep was instantaneous, even when he was in Moscow and she in St. Petersburg. The findings strongly suggest that the intention in the psychiatrist's brain was not sent in any conventional way, but that it found its target using some other vehicle. That vehicle, I believe, is what modern physicists call entanglement.

Entanglement

There are two good reasons why brainwaves, at least within themselves, do not constitute the primary elements of the miracle process. We get closer to the truth if we stop conceptualizing the process in terms of what gets *sent* from the influencing agent to the target and focus instead of how that agent becomes one with the target. A model for how that might work centers around a revolutionary concept in modern physics—the concept of entanglement. Dozens of experiments in what is known as quantum mechanics indicate that once two subatomic particles (e.g., photons or electrons) have collided at high speed, any changes taking place after that in one will automatically affect the other, even when the two are miles apart and no further physical contact between the two has been made. Because of the collision, they now act (at least in some respects) as a single entity even though there is no visible interaction between them. This would appear to contradict a basic premise in Isaac Newton's physics—namely, that even when they have interacted in the past, two objects that presently occupy separate spaces and have no physical contact with each other lack the capacity to influence each other directly. This is not only the old physics; it's what most of us assume to be the nature of the everyday world around us.

Once established at the subatomic level, the entanglement principle can serve as an analog of what happens in the world of objects, that is, the world as we humans know it through our senses. Assuming the analogy to be relevant, the implications are vast. As a result of being squashed into a pea-sized ball some 13.8 billion years ago, all objects that have evolved from that pea are physically entangled—some no more than tangentially (objects of different nature separated in time and space), others more intimately (e.g., identical human twins). As they were drawn into this pea-sized ball by gravity, the trillions of tiny, invisible particles that existed there collided with each other and in the process became entangled. Thanks to gravity, over the next 13.8 billion years those same trillions of particles were shaped into gases like hydrogen and helium as well as chemical elements such as carbon, iron, sulfur, and phosphorous. In time those basic elements combined to form stars and planets—some with water, an atmosphere, moderate temperatures and, given enough time, living creatures.

Hard as it is to believe, this means that we are in some way entangled with everything else in the universe—animate and inanimate alike. If entanglement at the human level has characteristics analogous to those operating at the subatomic level, people alive today (who are evolved from particles present at the Big Bang) should behave in a similarly entangled way. Recall that in the physicists' experiments, once particle A and particle B collided, any changes in one of the particles resulted in similar changes in the other, without any further contact between the two. At the human level, a somewhat different but related scenario could involve person A using her imagination to produce a desired change in person B. In such a case there are two assumptions required: first that person A is capable of creating and sustaining a sharp image of person B, and secondly that the entanglement between persons A and B is not only known conceptually by person A but is felt as a form of experience. If both assumptions are met, person A (via meditation) will gradually fuse with her image of person B, setting up the possibility of exerting influence from a distance—such as curing person B's cancer.

If valid, this scenario suggests that we humans theoretically have the power to influence any object we put our minds to.

We should, for instance, be able to read someone's mind even if we never met that person and live in a city 1,000 miles away. In that case, person A's image of person B would have to come from memory, e.g., from seeing B's picture in a magazine or on television.

If all of this is true, you will probably ask why so few of us have been able to influence others at a distance using nothing other than our minds. In his book Entangled Minds, Dean Radin suggests that it has a lot to do with the limits of our awareness.[iv] He writes:

> There are an astounding number of events we can potentially react to, but the vast majority of them can be regarded as background noise. Other than where your body is, you might be interested in perhaps ten other locations or events within the universe at any given moment, all of them relatively close to you in spacetime… Some portion of your *unconscious* mind pays attention to those selected locations at all times. [But] most of your *conscious* awareness is heavily driven by sensory inputs. That sensory-bound brain is also entangled and influenced by the rest of the universe, but its local effects are so much stronger and immediate than our "background" awareness that only on rare occasions are we aware of its entangled nature.[v]

Radin is saying that yes, you are entangled with everything else in the universe—and that includes other people, animals, oceans, trees, cars and stars, but because you are so completely preoccupied with your everyday life, you are unaware of these entanglements and rarely use them to influence others mentally. But that could change. If he's right, it means that by clearing your mind of all this "chatter", you can take advantage of the enormous potential that lies within you by virtue of your entanglement with the rest of the world. And that includes other people. If you want to cure a distant friend's cancer, for example, you don't have to be in his physical presence; you don't even have to send him healing energy. Because you and he are entangled, you can draw him to you with your imagination. All you have to do is create an image of the target person, then focus on that image to the exclusion of everything else. When that practice is sustained for a few minutes, it will bring him to you.

# 3

## How Miracles Are Made

As implied by the previous paragraph, the imagination plays a critical role in the process of making miracles. Englishman Neville Goddard put it clearly in a piece he wrote 75 years ago:

> When man solves the mystery of imagining, he will have discovered the secret of causation, and that is: Imagining creates reality.…An awakened imagination works with a purpose. It creates and conserves the desirable and transforms or destroys the undesirable…No object is independent of imagining on some level or levels. Everything in the world owes its character to imagination on one of its various levels…The world in which we live is a world of imagination, and man—through his imaginal activities— creates the realities and the circumstances of life; this he does either knowingly or unknowingly.[vi]

He's not only saying that your imagination plays an important role in turning desire into reality, but that without it you'll never get what you want. For Goddard this is a proposition not open to discussion. To attempt to change circumstances before you change your imaginal activity, is to struggle against the very nature of things. There can be no outer change until there is first change in one's imagination.

Goddard was writing in the early part of the 20ᵗʰ century and thus had no knowledge of quantum physics. Given that we now possess that knowledge, it is possible to link miracles not only to the imagination but to entanglement itself. When I say that, I'm not thinking of a cognitive link, i.e., a tie between two concepts, but an actual *experience* of entanglement. Getting to that advanced stage of meditation is a two-step process which starts with bringing your mind to a focus on the target (animate or inanimate) with whom you are presumably entangled, i.e., *isolating* that target to the exclusion of any other thought or feeling. Once you have the target isolated, you proceed to the next step where you *penetrate* the target, i.e. take your concentration to the point where you actually *experience* your entanglement with the target.

Once you get there, your life will never be the same again. When you get up from your cushion and begin moving around, if you remain focused on whatever forms you presently see, hear, taste, touch or smell, your perceptions will be possessed of a new dimension. While we have no commonly-used word for that dimension, I would call it a Oneness in which everything you perceive takes on a transparency in which things appear related both to you and to each other no matter how different they appear on the surface. In this new state of mind, other people, cars, trees, buildings, sidewalks, and even the nearby river appear united in a way that previously your sense organs were unable to pick up. But here's the paradox—despite the Oneness that pervades everything surrounding you, at the same time those forms remain distinguishable from each other. For instance, the bank on one side of the street and the grocery store on the other continue to be seen as different from each other (you don't walk into the bank and ask the teller where the tomatoes are), but they now "reveal a relationship which does not obtain in the ordinary field of vision."ᵛⁱⁱ

Later in the day you drive to Denver and head for the poorer sections of the city. Now when you walk the streets, you see high-rise apartment complexes; on the sidewalks you pass Hispanics, Blacks, Asians amid the row of tents serving as home to those who can't afford to rent a room elsewhere. You are newly aware that you are entangled with everyone and everything you perceive—and that includes people you can help with your recently discovered powers. While you consider the possibilities, the streets fill up with traffic, the noise is deafening, the air thick with noxious gases. There is hardly a tree, bush or flower to be seen. As you walk, you stay concentrated, knowing that if you do, ideas will occur to you for making the lives of those living here a little easier. Perhaps you can provide material improvements for their homes or disease-fighting tools that allow inhabitants to live longer, healthier lives—or even lives free of fear and violence. You leave but you cannot forget what you have seen. The prospects of what is possible keep bubbling up as you navigate the highway on your way home.

Once you have gained access to the experience of entanglement, the table is set to begin performing miracles involving the forms you are presently aware of—as well as those you can focus in your imagination. But first, you need an intention to influence the chosen target in a certain way. It could be an intention to cure someone's illness or to locate a stolen car or to lift yourself into the air.

Let's take levitation as an example. You begin by concentrating on the invisible force that's keeping you stuck to the ground, namely gravity. Even though you know it is a force powerful enough to have shaped the universe, you remind yourself that it is a presence you are entangle with—and thus a force which can be influenced. Once you have isolated that force in your mind, you summon the intention to defy it, i.e., to nullify (or at least modify) its immediate effect on you. You do that not by destroying it (which is impossible), but by neutralizing its impact on you—using the extraordinary power you have gained over years of meditation. Using nothing more than that power, you deny gravity its force, making levitation possible.

Goddard talks lot about imagination and intention, but at no point in his collected essays does he attempt to explain how they function in the paranormal process. From my own experience, I would say that the key consists of many steps, first clearing your mind of all distractions so that there is little or no separation between you and your image of a target. After doing that daily for a few months, you may experience a slight shrinking of the self–a change that reveals itself in less bragging, less defensiveness, and less concern with protecting your self-image. Even a modest shrinking will offer the opportunity to exert influence on whatever target you're focused on. For example, you may get an intuitive sense of what that stranger sitting next to you on the subway is thinking—your first hint of the paranormal, in this case telepathy. While this is encouraging, it is just the beginning. As your meditation deepens, the mental gap between you and your target image will narrow, whether that target be a person or an inanimate object. In the process, your awareness changes from seeing yourself as an outsider who is observing an image of something else—to a state where you as observer disappear, leaving nothing but an image of the target. At that point there is no observer left, no witness, no inner "I". You have lost yourself in the image of the target.

Although I have no evidence to support the idea, my guess is that as your practice deepens, your awareness will progress along a perceptual gradient which begins with a clear-cut distinction between observer and observed and ends (if you take it that far) with the absence of any distinction between the two. If that is what happens, it follows that your location along that gradient will affect how successful your interventions are likely to be. If, for example, you are still having trouble concentrating on your target, you will probably fail when you try to heal your friend's sniffles or intuit what she is currently thinking. Years later when your concentration is not only much sharper but sustainable over longer periods of time, major miracles like walking on water and changing the weather become possible. It should be added that such events are also possible (although rare) among individuals who have never meditated but were fortunate enough to inherit the right combination of genes at birth.

To summarize, there are several distinct steps in the process of creating a miracle. First you bring your meditation to a white-hot pitch by clearing away all distractions and concentrating on a single thought (e.g., a mantra), word (e.g., MU), or sensation (e.g., your breath). When your mind is perfectly clear, you summon an image of the chosen target and hold that image until you are conscious only of two things—the target and yourself as observer. Next, the awareness of yourself as observer must go. When that happens, there is no "you" left in awareness—nothing remains but the image with which you are entangled. Keep in mind that you have not created the entanglement; it has been there since you were born but because it lies buried under a mountain of everyday concerns, plans, emotions, and memories, you haven't been aware of it.

You back up now and recall how you want to influence your target. Because you are re-entering the realm of language and forms, you will experience a slight loss of intensity, but it can't be helped. Your power is still strong enough to influence the target remotely. This is true whether the target is another person's cancer you want to heal, a roulette ball you want to manipulate, a hurricane you want to deflect or a stolen car you want to retrieve. Through it all you keep your mind focused on the target until the process has been completed. You will know that you have reached that point when the image before you manifests the desired change. Depending on circumstances, you check later to see if the actual target has undergone the same change as the image.

If the act of creating a miracle still seems murky, it might help to view the process like an archeologist. First, you clear away the rocks, soil, and vegetation (*akin to distractions*) at the surface of the site. As you do so, you carefully sift through the contents for signs (*akin to hunches, premonitions, intuitions*) of an ancient civilization thought to be buried underneath. After much digging (*meditation*), you come upon unmistakable evidence of a lost culture (*awareness of entanglement*), one that presumably has been concealed for centuries by the soil, rocks, and trees. You run your hands over the artifacts left by these ancient people and gently bring them to the surface (*consciousness*) where you can study them. In time the results of your investigation are published in professional journals. The dig fulfills its purpose when those findings expand the world's awareness of how closely we are linked to cultures that appear to be distant from us in both space and time That new awareness allows us to resurrect the lives and structures of a world we once thought lost forever (*mental influence at a distance*).

# 4

## *The Self as Obstacle*

So far, I have argued two things, first that the human brain is extremely powerful, far more powerful than it was previously thought to be—and secondly, that it takes a lot of mental energy (the kind that typically comes from years of meditating) to create a miracle. Assuming there is more than a grain of truth to these two arguments, you might ask why miracles are so rare. Afterall, with a brain as powerful as I say it is, there must be at least a small but significant number of advanced meditators who have demonstrated the ability to do miraculous things, like manipulating a ball with the mind alone, changing the weather or walking on water. Thus far there is no scientific evidence that such things are even possible. So, what's going on? If our brains are as powerful as I say they are, why haven't more of us used that power to perform miracles? If the power is potentially there in all humans, what's stopping us from using it?

Theoretically, years of meditation should generate enough power to produce a miracle—but that hasn't happened. It may not be enough because there is something—something equally powerful standing in the way. I refer to the self, the inner "I' with which we identify, i.e., the person we take to be who we are. This is the inner "I" we (wrongfully) assume that is guiding our choices and making our decisions. This is the "me" we love above all else in life. But it is only a flattering illusion, one spun out of language and the misuse of personal pronouns. It may be nothing more than an illusion, but over the centuries it has cut us off from our connections both with each other and with nature as a whole.

While I have been meditating for decades, it is only in the last seven or eight years that I have taken my practice seriously enough to elicit thoughts of the paranormal. Such thoughts have been prompted by an awareness that what I experience in meditation has gone deeper. Because it is free of most distracting thoughts and emotions, my daily practice feels more intense and, due to that intensity, more powerful. It is as if I had entered a different realm of being, one no longer centered on me but rather one without any center at all—a world without boundaries, self, or objects— and for that reason a world beyond description. And that raises a profound question about my identity.

If there is no "me" in this new realm of being, do I even exist as a person—as a self, an "I", as a separate individual? At the deepest level, the answer is no. In this new realm, "I" am no longer a "who" but rather a "what". As Peter Ralston puts it, I am "nothing in particular." There is no inner agent here that exercises control over my thoughts and behavior— not even an inner "I" that is *having* these experiences. Because I am undifferentiated in any way, the only thing that can be said about me is what I am *not*. I am neither subject nor object, neither observer nor observed; I have no locations; I have no time; I have no boundaries; I have no color; I have no beginning or end. So, what am I? I am the empty space that abides here. Objects of perception (things that can be seen, heard or felt) arise in this body-mind when prompted by events, but leave when they are no longer needed. At all times the empty space (the real "me") remains, either in the background when objects are being perceived or in the foreground when they are not (e.g., during meditation). If a new sense of power is arising here, a power so strong that it makes miracles possible, it owes its provenance to the discovery that this empty space is "what" I really am—and have been since birth.

So, what is preventing us from using this power to perform miracles? The answer lies not in the physical brain itself but in the consciousness to which it gives rise. As suggested, we humans (unlike our red-in-tooth-and-claw cousins) typically identify ourselves with an internal agent (the inner "I") that serves (we believe) as the master controller behind most of our thoughts and actions. That belief varies across cultures, but is held most strongly in Europe and the United States. Members of Eastern cultures like Japan and China still identify with an inner "I", but define themselves just as strongly in terms of family (Japan) and nation-state (China).

Despite what psychologists say, trouble can begin with any sort of identification, but is particularly problematic in those societies where the inner "I" (also called the self, ego or soul) is taken as one's basic identity. In those societies children are taught at an early age not only to define themselves as an inner "I", but to hold that agent (in other words themselves) responsible for much of their behavior. And that's where the trouble begins—the belief that it is neither our genes nor our upbringing that determine our behavior (which is the truth) but the inner "I" exercising its innate, contra-causal power (which is an illusion).

In turn, that belief affects how society deals with social deviance. In the Western world where identification with a free-willing inner agent is most strongly established, the primary defense against those who break society's norms is guilt; in the East where one's identity is strongly shaped by family and country, defense against deviance is more apt to take the form of shame. Besides guilt and shame, of course, every society has its own unique set of public penalties for controlling deviant behavior (e.g., fees, withholding of licenses, imprisonment, torture, and banishment).

It was the early Buddhists who contended that there is no such inner agent in any culture—in other words, that the inner self that almost everyone identifies with to some degree is an illusion. Some analysts have suggested that the "I" that we falsely identity with is the source of most human suffering. That is not as big a leap as you might think.

Consider how central a role identifying with a free-willing inner agent plays in getting along with others in today's highly interactive world. A century ago, most of us were either farmers working alone in the field or mothers who stayed at home cooking, cleaning and taking care of children. Today most of us (men and women alike) work in organizations—factories, hospitals, schools, companies, stores, clinics, law firms, universities or government bureaucracies. This represents a huge change in the kind of experiences we have. In such a highly interactive environment, identifying oneself as an inner "I" that is causally responsible for what it says and does leaves one exposed to constant evaluation by one's peers. And that can be a problem.

Believing that one is a free-willing controller of thought and action may give one a sense of power but that power comes at a price. The price is tension, the kind that Thoreau had in mind back in 19th century Concord when he said that "Most men lead lives of quiet desperation." While he stopped short of linking that tension to a belief in free-will, he made it clear that it was due to the way we lived— and that we would be better off leading a simpler life.

If he said that almost 200 years ago, just imagine what he might say today in this world of radio, television, cell phones and the internet. In this world there is no escaping how others respond to what we say or do— and given our identification with an inner "I" that is causally responsible for our actions, that is a recipe for tension. We deal with tension in different ways, some with cigarettes, alcohol, or drugs—others with pills, withdrawal, overeating or denial. In any case, persisting tension generates anxiety. We worry about what others will think; we fear their disapproval or outright rejection. And this stirs up thoughts and feelings that drain energy from the brain, leaving us helpless to realize the potential that those brains are designed to offer. But there is reason for hope here. With sustained meditation we can sweep away these tension-inspired thoughts and feelings and gain access to the enormous power of the brain. Doing so will not only relieve our anxiety; it can even make us masters of the miraculous—an exalted status once reserved for the gods.

Identifying who and what we are with an illusory inner "I" that can be blamed for its choices gives rise to a level of anxiety that makes it impossible to harness the full power of our brains—in particular, the power to perform miracles. By itself that is enough of a problem to explain why miracles are so rare. As stated earlier, there is something else at work here—something hidden from view by the belief that we are free-willing, autonomous individuals isolated from both each other and nature. But as modern physics has made clear, we are not separate after all. Regardless of what appears to be the case on the surface, we are entangled with everything else in the universe. To be entangled with other humans, animals, planets, stars, rivers and oceans means more than just being connected in the ordinary sense. We are already aware that we are connected to other humans geographically; we pass them or may even stop to talk to them as we walk down the street. We are connected as well to the sidewalks we walk on, to the buildings we pass on our way, even to the street lights that blink on and off at the intersections. And of course we are connected to family members, friends, neighbors and people we work with. But being "entangled" with other entities means more than being connected in any of these obvious ways.

But here's the rub. Identifying with an illusory self has up to now prevented the process of miracle-making from playing out by anchoring our identity in an illusion which defines us as separate from everything else in the universe. And that includes all animate and inanimate forms here on earth. But as research in entanglement suggests, we are not "other" than the birds we enjoy watching or the dogs and cats we love to pet. We are not even "other" than the sun that keeps us warm or the rain that makes our vegetables grow. We "co-exist" with them all. We may be dramatically different in looks or the genes we are born with, but we are not separate. We continue to believe that we are separate because the way we identify ourselves prevents us from seeing that at our core, we are one with the sub-atomic particles from which we originated and out of which we are presently formed. It is those basic particles that entangle us with the rest of the universe and in the process make all kinds of acts possible, including the ability to perform miracles. The secret to regaining that power (which we once had but lost during the evolution of our species) lies in awakening to the truth that at the deepest level, we are connected to everything else in the universe.

# 5

## *Cultivating the Paranormal*

While there are many techniques that can be employed in cultivating the ability to perform miracles (chanting, dancing, fasting, etc.), meditation remains the primary means of developing such power. Sustained meditation clears the mind of all distractions, allowing you to concentrate attention on a single thought, desire, image, or intention. Once all distractions have been eliminated, it becomes possible to make changes in the entangled target you have focused on, whether that target is internal (e.g., your own physiological state) or a remote object (animate or inanimate). Meditation, however, does not *create* entanglement nor does it improve the entanglement that existed at the time of the Big Bang. What it does and does very well, is to clear the mind of distractions, which allows the entanglement to reveal itself.

When we use our minds to influence someone remotely (e.g., curing someone's disease), we typically conceptualize it as sending something to a target (a message, healing energy, etc.). This is a mistake. It would be more accurate to think of it as drawing the target *to* you. But you don't have to bring the actual target physically close; all you have to do is *imagine* the target being close to you. And that is done by forming an image of the target while clearing your mind of all competing thoughts and sensations. The more vivid and precise your image, the greater will be your power to influence the target by replacing that image with a more positive one, for example, by replacing the image of a diseased organ with the image of a healthy one. The basic act needed for advancing the process is your intention to influence the target in a particular way. Thanks to entanglement (and a clear mind) that intention will eventually bring about a change in the actual target.

If increasing the mind's ability to concentrate is the primary way to cultivate the power to influence events remotely, the major obstacle to that sharpening is our identification with an inner "I" that demands to be seen in a positive light, both by others and by ourselves. As was said previously, that inner "I" thought to dwell somewhere inside our bodies doesn't even exist; it is nothing more than an illusion sustained by our culture, in particular our language. Our obsession with promoting and defending that illusion keeps us in a near-constant state of what Henry Thoreau called "quiet desperation." That obsession demands that we constantly remain on guard against the possibility of being frowned upon, criticized, forgotten, passed over or outright rejected. And we needn't be at work or in a relationship to have those crippling thoughts and feelings arise. They stand ready to invade our peace even when we are alone; all it takes to let them in is a little fear and self-doubt. And we typically have plenty of both.

It is that state of anxiety, fear and worry that must be eliminated if we are to gain access to the potential power of the mind to shape the world around us. The elimination process is analogous to making maple syrup by boiling off most of the water contained in the maple tree sap. The sweet-tasting syrup is there to begin with but in its natural state, is hidden beneath the copious amounts of tasteless water surrounding it. In our analogy, the water diluting the sweetness is comparable to the thoughts and feelings that obscure the object of our concentration. Just as we bring the sap to its maximum sweetness by boiling off most of the water, meditation maximizes our mental power by eliminating our distracting thoughts and feelings. It is the undistracted mind, the mind that has been cleansed of all chatter, that holds the power to create miracles via entanglement with other entities.

But what about changing yourself first? As all religions testify, don't you have to undergo some sort of personal transformation before you can hope to change others? Don't you have to become less preoccupied with your own agenda and more concerned about others before you can gain the power and moral authority to remake the world? In Christianity you are called to imitate Christ (or at least make an effort in that direction); in Buddhism you are called to follow the Eightfold Path (right understanding, right speech, right action, etc.). But could both religions be wrong? From what I have learned through meditation, these fundamental improvements in personality should not be thought of as requirements to be met through deliberate effort, but as changes that emerge by themselves as you meditate. They come, in other words, as a by-product of cleansing the mind of debilitating distractions. They come automatically in response to dispelling the primary cause of those distractions, namely identification with an illusory inner "I", in other words the ego. That suggests that instead of embarking on a spiritual journey (even one centered on performing miracles) by flailing yourself as sinful and unworthy, you'd do better to work on bringing your mind to a single point.

There are two primary types of meditation currently in vogue, each with its own virtues. In mindfulness meditation the seeker adopts a receptive frame of mind and waits for a sensation, feeling or thought to arise. The idea is to allow mind states to arise spontaneously without clinging to any of them. If a particular thought or feeling enters consciousness, avoid the temptation to pursue it further. Acknowledge it (be mindful of it)—and then let go and return to a state of readiness for whatever arises next. If practiced devotedly, the result is an increased clarity of both thought and perception. As that unfolds, the mind becomes more peaceful, making life's problems easier to handle.

The other type of meditation requires more effort but is better suited to developing the power to create miracles. It takes more effort because the seeker is instructed to fix his or her mind on a single sensation, image, mantra, or koan while excluding awareness of anything else. The goal is simple—to train the brain to focus on one thing at a time (in a word, to perfect the ability to concentrate)—but the practice is anything but easy.

In Zen Buddhism there are two different schools of thought—Soto Zen where you concentrate on your breathing—and Rinzai Zen where you concentrate on solving a koan (e.g., "what is the sound of one hand clapping"). What the two schools have in common is a belief that concentration, when carried to the point of becoming One with the object of your attention, will lead to that state of undifferentiated emptiness that the Buddhists call enlightenment. Reaching and sustaining such a practice makes all things possible, including the performing of miracles, even though most adepts choose teaching as the preferred way to help others.

When sustained over time, concentration makes miracles possible, whether you pursue the deeper experience of enlightenment or not. More than anything, performing miracles requires an abundance of energy—and training your brain to concentrate on one thing at a time is critical for developing that energy. While mindfulness meditation is likely to bring peace to your life—and make it easier for you to accept the obstacles that life throws in your path, it lacks the power to unlock the energy that is currently being spent on self-preservation.

To make that energy available for miracles, it is necessary to shrink the self-illusion to the point where it no longer serves as a distraction. And the quickest way to do that is to bring the power of the brain to a single point through sustained concentration on a mantra, word or sensation.

Everything I have said thus far about meditation refers to the standard meaning of the term—the practice of sitting in a certain position and focusing your mind on a particular word or thought. However, there is another, less formal strategy that can add to what a traditional sitting practice offers. We start with the assumption that the key to accessing the paranormal is letting go of the illusory, energy-devouring inner "I". In the informal practice I have in mind (called "sense immersion") you focus on a sensation (e.g., as you eat, walk, make love, listen to music, etc.) to the exclusion of any thought about the self. For example, take the sound ( "crunch") associated with eating an apple. The technique is straightforward. Once you have that sound in your cross-hairs, burrow into it so deeply that you lose all awareness of an "I" that is taking part in the experience. What remains when you succeed in doing this is the "crunch" itself— no "I" making the experience happen, no "I" observing it, not even a "me" to whom the experience is happening.

If there is no inner agent making that experience of a crunch happen, you might ask how the experience originates. The short answer is that the sound is constructed by different parts of the ear and brain. It starts with an energy wave which originates when something vibrates, e.g. teeth biting into an apple, creating a collision with the air molecules around it. In turn that collision produces a wave of mechanical energy which is collected by your outer ear. In your middle ear, that wave is amplified and transmitted to the inner ear where a special organ converts it into neural impulses located in the audio cortex of the brain. It is at this point that we first become conscious of hearing something.

In other words, the sound of "crunch" starts with vibrations in the mouth that give rise to sound waves picked up by the eardrum and passed on to various bones and nerves in the inner ear and eventually to the brain. There is no role for an inner "I" at any point in the process. The originating cause of the experience is the impact of teeth on an apple and not the work of an illusory homunculus hiding somewhere in your pineal gland (as Rene Descartes would have it).

With this explanation in the back of your mind, you can use the "crunch" experience to further your grasp of the idea that nothing in consciousness (whether sensations, choices or decisions) requires the presence of an inner "I". Try this. As you burrow into the sound of "crunch", allow the thought to arise that there is no sense of an inner "I" accompanying that sound. You are experiencing what the Zen Buddhists refer to as "no-mind". Do this over and over, not just with sound but with a variety of sensations as well. When done consistently, the practice will cement both the knowledge and the feeling that the inner "I" (the self) is nothing more than an illusion.

Here's a question that seekers of the paranormal are likely to ask: Is it possible to give such goals the priority they require while remaining fully immersed in the ups and downs of everyday life? Regarding enlightenment, Tibetan scholar and explorer Alexandra David-Neel, would say no. Tibetans who aspire to enlightenment typically spend months or even years secluded in caves or huts free from the distractions of village life. The most serious aspirants even go for long periods (in some cases 5– 10 years) without any physical contact with outsiders. They survive thanks to nearby villagers who climb steep cliffs to supply their adopted recluse with food and supplies while carefully avoiding any interaction— a major sacrifice (particularly in the winter) which is usually performed less out of compassion than in the hope of gaining merit for themselves and ultimately a more favorable rebirth.

This is the "hermit" model developed over the centuries in Tibet, but is it the one contemporary spiritual seekers have to use? Not necessarily. If you belong to a formal religion, there is likely to be a monastery or rural retreat where you can live and continue your practice. For individuals of a more secular bent, particularly those seeking to develop the capacity to make miracles, there is an alternative which is likely to sound more doable to modern ears, namely "Be in the world but not of it." Being in the world means not being a recluse in a hut tucked away somewhere in the hinterland, but that raises a serious question: how can you live in the world, surrounded by other people, subject to the demands of family and job, not to mention those of neighbors, landlords and tax collectors— without being a part of that world? You can't.

The secret is to live in the world and be a part of it without becoming *attached* to it. Being attached to something (whether human or otherwise) implies hanging on to it— often at great cost to your peace of mind. But what about love, you ask? It is at this point that most people protest, insisting that you can't love something or someone without being attached. The two are identical—so they say. Arguing that way confuses real love (what the Greeks call agape, meaning love for another without expecting anything in return) with the kind of love that is more familiar, namely the kind where we genuinely care for another but expect to get something for ourselves in the process.

To unburden yourself from attachments (at least those involving love of another human being), you can start by asking what you want from the relationship. Most people would say that at the very least they want to be loved in return. That implies that they're probably going to hang onto the relationship; in a word, they're attached—and will suffer the consequences of that attachment. If your ideal is to be *in the world but not of it*, your attachment to another person is going to make that impossible. So, how do you remain in the world without becoming attached to anyone or anything in it? If, for example, the other is a woman, learn to love her without hanging onto her. When you are together, you do that by focusing on *her* needs and putting your own needs aside. In that way you maximize what you *give* to her and minimize what you want *from* her. You remain in the world but are not fully of it. If something horrible happens and she dies unexpectedly, you focus not on how much you are going to miss her but on what a lovable person she was and how lucky you were to have been part of her life.

The unattached way of relating to your children is not that different. You don't insist on being loved in return; you don't groom them to do something with their lives that will make you proud or fulfill some unrealized ambition of your own. You teach them the essentials and then step back and do what you can to support them as they give shape to goals of their own choosing.

With inanimate objects like homes, cars, stocks, jewelry and cash it's different in some ways and similar in others. It's different in that it's a one-way relationship; you can't expect your car or jewelry to love you in return. But being attached to some of those inanimate things, like a home or car, can be similar to being attached to a person in that they involve clinging. And that means a sustained fear of loss. So, what can you do? Assuming that you aspire to serving others, you remind yourself that if you buy a house, there is no guarantee you will have it forever. You can't possibly know now what kind of weather is coming in the future, or what the local economy will do, or what financial decisions you will have to make in the coming years. With all those unknowns in mind, you accept the possibility that if you buy the house, you may only have it for a short time. More broadly, you stand ready to accept whatever happens in the future, leaving the possibility open of making changes elsewhere in your life if the house is damaged or has to be sold.

But there is something else going on as you struggle to accept events whether you like them or not. Once you decide to include meditation as an integral part of your life, the hold that objects like a home or car have on you gradually lose their grip. Your desire for such objects, once inviolate, begins to fade. More broadly, you begin to see all events (not just objects) as coming and going, arising and falling, each the inevitable outcome of the events that preceded it. And that wider perspective makes the loss of any objects you are attached to easier to accept. As your mind is cleared of inner chatter (most of which centers around your ego), your desires shift from things you want for yourself to those that will benefit others. You spend less time thinking about your home, your car, your jewelry, your bank account, or your reputation and more time pondering how you can reduce suffering in the world. In the process you become more and more self-forgetful—a change which if serious enough, will catapult you to the point where you become capable of paranormal powers.

For most modern people who are serious about performing miracles, "being *in* but not *of* the world" is the path most likely to be chosen. But it is important to recognize that such a choice comes with its own set of problems, first and foremost among them being time. Although no one to my knowledge has ever researched the subject, my guess is that it takes five years for the average Tibetan cave-dweller to reach enlightenment: by contrast, someone who hopes to achieve a similar goal by "being in the world but not of it" will require at least double that amount of time. A similar ratio probably holds for those looking to perform miracles. The reasons for the time difference should be obvious. The monks who live in a cave can spend as much of the day as they choose to meditate—and do it without interruption. Their opposites have no such luxury. Part of the problem comes from the situation in which the latter find themselves—the demands of a full-time job, the needs of a spouse and children, and the obligations of being a citizen like paying taxes, managing a bank account, getting your car inspected, repairing your house, recycling waste products, not to mention putting aside enough money so your kids can go to college when they grow up. And then there are all the things you don't have to do but are likely to do anyway, like watching TV, listening to the radio, talking to friends, chatting with neighbors, planning and taking vacations, buying birthday and Christmas presents, writing thankyou letters, getting an annual health checkup, deciding whom to vote for, reading spiritual books, and taking the dog for a walk.

You might object to some of the items on the "optional" list, but some of them are necessary if you want to make changes in the world. You can't make serious changes in world politics, for example, if you don't know what's going on nationally and internationally. You not only have to be aware of what's going on—you have to give it serious thought. The same is true for local programs like those that help the poor, the homeless, the disabled, and the marginalized. All that thinking is likely to fill your mind with concerns that cloud the mind and prevent you from accessing the paranormal. More simply, all that cognitive activity keeps you from meditating. And that's why "being in the world but not of it" takes so much longer than living in a secluded place where you are basically cut off from the world. But there's a price to be paid for choosing the hermit's path. When you finally come out of your cave, your lack of knowledge about the world limits what you can do to serve others. Time has moved on without you and you may never be able to catch up. You may have developed the capacity to perform miracles, but whether you can use that power effectively remains uncertain. You may not even want to.

# 6

---•◦●◦•---•--- ---•---

## *Possible Misuses of Remote Influence*

The extraordinary power that comes with intense concentration would appear to invite the possibility of malicious behavior. For example, it might be possible (although there is no evidence to support this) to cause a heart attack in someone just by dwelling on the wish mentally. All kinds of mayhem could arise as more and more people cultivate the brain power required for such an outcome. But there is good reason to believe this is unlikely to happen. For one thing, not that many people have the persistence and mental stamina required to develop that kind of power. But even among those who do, there are built-in safeguards against using one's power maliciously. For one thing, paranormal abilities can't be developed without a significant shrinking of the inner "I"—that illusory inner agent that typically responds to status threats with hostility and a desire to harm. The shrinking of the self that arises with intense meditation should act as a safeguard against the arising of such negative feelings. As suggested earlier, it promotes just the opposite, namely feelings of tolerance and compassion.

So, what does all this have to do with the concern that the pursuit of paranormal powers might lead to malicious behavior? It is unlikely to do so because the meditation required for that pursuit has the effect of loosening and ultimately erasing the identifications that lie at the heart of human conflict. As that meditation deepens, the individual self and its accompanying identity begin to shrink, leaving less of an inner "I" to identify with. As a result, you not only spend less time thinking about yourself, but are less concerned about advancing and defending whatever it is that you identify with. And that implies being less likely to use gender, race, social class, religion, or nationality to define who you are fundamentally. And for that reason, you are less likely to become defensive when your gender or race is disparaged, your social class belittled, your religion attacked or your patriotism challenged.

Once the psyche is freed from attachment to identity, it becomes imperturbable; like the deep ocean, it remains calm no matter how turbulent events get on the surface. When we wake up to the fact that what we really are lies much deeper than the way we typically define ourselves, success and victory as well as failure and defeat can all be taken in stride—and that shift is guaranteed to bring in its wake a freedom unknown to those still clinging to a self-image. The secret to a more peaceful and joyful life is to stop taking the inner "I" seriously. And that can be done while living *in* the world as long as you remain unattached to anything or anyone in it—including all the things you identify with.

# 7

# An Historical Perspective

It is apparent from myths, cave paintings and tomb inscriptions that our earliest ancestors were completely embedded in nature, much like the animals they hunted. As such, their conscious lives were guided by instinct and impulse rather than language and logic. In place of abstract concepts and reason, paleo consciousness was filled with emotions, impulses, images, premonitions, and remembered dreams. While their brains were as large as ours, their thinking and feeling relied less on the prefrontal cortex and more on the amygdala. That kind of brain activity placed severe limits on their rational thinking but was compensated by a greater capacity for what we call the paranormal. Assuming that to be true, it follows that they must have been far more expert than we are today in reading minds, healing the sick, seeing into the future, and communicating with animals.

As human culture evolved, however, all that changed. Increased success in adapting to the natural environment through farming led to a gradual differentiation from nature. Evolutionary biologist L.L. Whyte put it this way:

> The attention of the individual was drawn more and more to his own thought as well as to external stimuli, and he became aware of himself as a thinking and feeling person endowed with the faculty of choice.[viii]

All this happened in Mesopotamia around 1500-1000 B.C when according to Joseph Campbell, the individually-minded shepherds of the hills swooped down to the cities below, crushing the urban dwellers and their collectivistic ways. What they brought with them was a new culture in which individual freedom was given priority. But, as Ken Wilbur reminds us, "it is one thing to gain a freedom from the fluctuations of nature, emotions, instincts, and environment—it is quite another to alienate them. In short, the Western ego did not just gain its freedom from the Great Mother; it severed its deep connectedness with her."[ix] According to Wilbur, it was then that mankind broke free of "subservience to the Chthonic Mother— and established itself as an independent, willful and constellated center of consciousness, a fact represented in the Hero Myths."

And that was the beginning of a belief in free-will. For the better part of the last 3,000 to 4,000 years humans (particularly those in the West) have been tricked into believing that having choices means that at least some of the time we can act independently of previous conditions—that we have the unique power to make things happen without being at the effect of anything else. Breaking free from the Great Mother's suffocating embrace was clearly a good thing; it gave us the power to think for ourselves, to pay attention to our own desires, and act independently of the tribe, but in the process of asserting our freedom, we went too far. As L.L. Whyte puts it, "We not only *differentiated* ourselves from nature—we *dissociated* ourselves from it."[x] We cut the bonds that tethered us to our environment; in the process we convinced ourselves that we could bend life to our will, that we could impose our intentions on reality even when conditions were adverse. We created heroes we could worship and learned to define ourselves in their image. In our ecstasy, we transformed ourselves into minor gods.

But this is an illusion. We are not minor gods. We certainly have choices, but those choices are not causally free. We don't have the power (usually) to choose without being at the effect of previous conditions. Our choices, thoughts, feelings and behavior are shaped by the genes we inherited, by the way our parents treated us, the culture we were raised in, the technology we used, the friends we made, the people who taught us, the books we read, the news we watched on television, the food we ate, the diseases we endured—and so on. But to most people, whenever we are confronted with a need to decide an issue or make a choice, it still feels like we are free to do as we please. That is what our culture teaches us—even as it assumes the presence of an interior controller for which there is no evidence.

In insisting that we have the power (at least some of the time) to act independently of both biological and environmental forces, we have divorced ourselves from the rest of the world— from our surroundings, from nature, from the circumstances that have made us what we are. And that is an illusion. It was fine for our ancestors to wriggle free from a vegetative existence where decisions (when to plant seeds, when to harvest, etc.) were made not by individual farmers but by priests and sorcerers. Rebelling against this hieratic order, however, came at a price. By creating the illusion of a free-willing controller tucked away inside of us, it left us identified with a fiction we felt obliged to defend and preserve—and that fiction remains to this day the source of most of our suffering.

Our progress to date has been marred by costly mistakes. While we no longer sacrifice our young to the gods in hopes that assuaging their thirst for blood will persuade them to send us good weather for our crops, we haven't given up our belief in the power of divinity. By redefining ourselves as contra-causal agents, i.e., as individual selves with the power to cause events without being at the effect of other causes, we have instead turned into gods ourselves—minor gods to be sure but deities of a sort nevertheless. If it's not already obvious from your personal experience, be reminded that in the evolution of human consciousness this ego-saturated way of identifying ourselves has been the primary source of our suffering.

But another error of equal significance was soon to follow. As our ability to reason and conceptualize progressed, our familiarity with the paranormal fell into disuse. Paranormal events, now increasingly rare, were attributed to the actions of the gods alone. If someone cured the lame or walked on water, it was because he was either a god himself or was affiliated in some way with the divine. In the light of reason and logic, such acts were seen as lying beyond the capacity of mere mortals. And with a few exceptions that's the way it remains today. While we might muster an occasional healing or even predict the order of a small deck of cards, the big miracles are still off limits to humans.

There are two contrasting beliefs here, both erroneous. In the case of free-will we have attributed to ourselves a power traditionally reserved for the gods, namely the power to control events through the sheer exertion of will. In making the gods sole enablers of the paranormal, we have at the same time committed the opposite mistake of attributing to supernatural forces a power which is really our own. To correct the first error, we must give up our identification with an illusory, free-willing inner "I"; to correct the second, we have to acknowledge that it is we (not the gods) who have the ability to perform miracles. The two corrections go together; giving up the illusion of a free-willing inner "I" will clear the mind of distractions and make access to our hidden paranormal powers easier. In the end, correcting the two errors will leave us with a more accurate image of ourselves—more humble on the one hand, more powerful on the other. To bring that about, we need a trade-off; we must give up identifying with a free-willing, contra-causal controller that doesn't exist (an act that will make us feel more humble) while at the same time reclaiming from the gods a power that we once shared (an act that will make us feel more powerful). The result of the project will be a radical overhaul in the way we perceive ourselves and our relationship to the world. And that is a project worth pursuing.

# Endnotes

i. Radin, D., *The Conscious Universe*, HarperCollins, New York, 1997, p. 324, italics added)

ii. David-Neel, A., *Magic and Mystery in Tibet*, New York, Dover, 1971, p. 291

iii. Personal communication

iv. Radin, D., *Entangled Minds*, Simon and Schuster, New York, 2006, pp. 264- 265).

v. *Entangled Minds*, p 264

vi. Goddard, N., *The Essential Collection*, Merchant Books, 2015

vii. Herrigel, E., *The Method of Zen*, New York, Vintage Books, 1974, p. 46

viii. Whyte, L.L., *The Next Development in Man*, quoted in Wilbur, K.., *Up From Eden*, Shambala, Boulder,chs. 11 and 12.

ix. *Up From Eden*, chs. 11 and 12

x. Campbell, J., *Occidental Myths*, Penguin Books, New York, 1964, p.24.

www.ingramcontent.com/pod-product-compliance
Lightning Source LLC
Chambersburg PA
CBHW040910210326
41597CB00029B/5037